UNDERWATER

Diane James & Sara Lynn

Illustrated by Siobhan Dodds

2   Introduction

4   Whales

6   Starfish

8   Octopuses

10   Sea Anemones

12   Dolphins

14   Crabs

16   Strange Fish

18   Seahorses

20   Sea Turtles

22   Quiz

24   Index

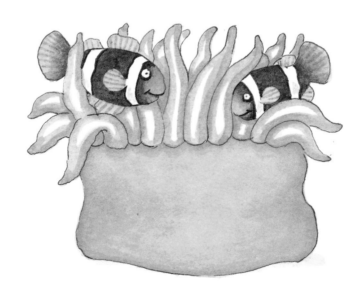

TWO CAN ™

PRINCETON ▪ LONDON

Have you ever been swimming? Moving about in the water is quite different from moving on land. Sea creatures move, eat and breathe in different ways than land creatures.

WHALES

Whales spend most of their lives underwater. They come to the surface to breathe through blowholes on top of their heads.

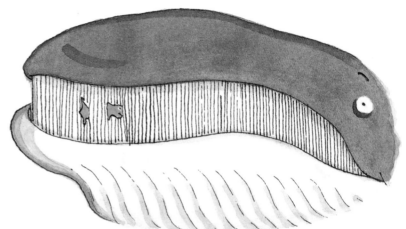

The blue whale is the largest animal in the world. It eats tiny creatures that it sifts from the water.

Whales swim by beating their tails up and down in the water. Fish move their tails from side to side.

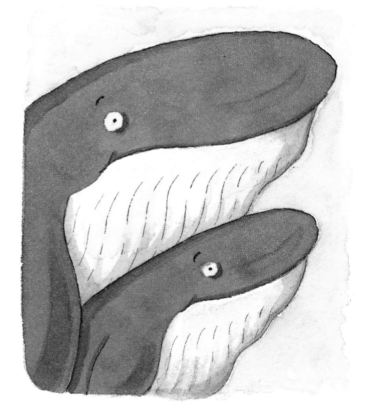

Baby whales stay close to their mother when they are born. They feed on her milk.

STARFISH

Most starfish have five pointed arms with rows of suckers under them. A starfish has its mouth underneath its body.

Starfish use their suckers to creep along the sea bed, to breathe and to pass food to their mouths.

Some starfish have lots of arms. They are known as 'sun stars' because they look like the sun and its rays. Starfish are usually very brightly colored.

Many starfish eat shellfish. They wrap their arms round the shells and pull them open.

OCTOPUSES

Octopuses have soft bodies with no bones. They have eight arms which are called tentacles.

Octopus tentacles have rows of suckers which help them to grip on to rocks and catch their food.

The octopus sucks in water and squirts it out through a hole in its head. This pushes the octopus through the water.

Octopuses can change color when they are excited or when they want to hide.

SEA ANEMONES

Sea anemones look like brightly colored underwater flowers, but they are really animals in disguise!

If you disturb a sea anemone, its tentacles fold away. They also fold away at low tide so that the anemone does not dry up and die.

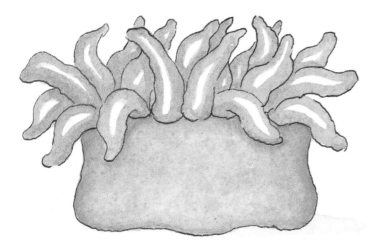

The tentacles of a sea anemone look like the petals of a flower. They can be used to sting small fish and pass them into the sea anemone's mouth.

Sea anemones spend their lives attached to rocks. They can only move very slowly. But sometimes they somersault to a new spot!

DOLPHINS

Dolphins belong to the same family as whales. They come to the surface of the water to breathe.

Dolphins have their own language. They squeal, bark, click and even whistle to talk to one another.

Dolphins eat small fish and squid. They catch their food and then swallow it whole.

Dolphins are very friendly and playful. They love to roll in the water and leap out of it.

CRABS

Crabs have soft bodies covered by hard shells. Their eyes stick out on short stalks.

Crabs have ten legs. They walk sideways instead of forwards!

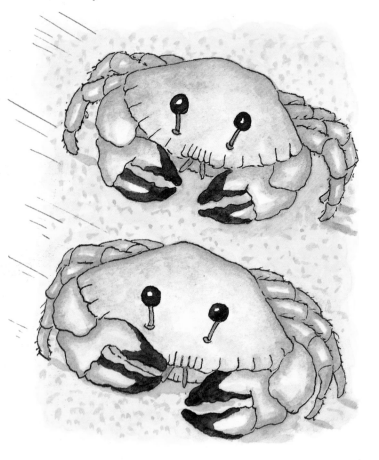

Crabs have strong pincers on their two front legs. They use these to protect themselves and to catch their food.

Some crabs use their back legs as paddles when they swim.

STRANGE FISH

The angler fish is an underwater fisherman. It wriggles a special fin like a worm to attract smaller fish.

Most fish avoid the stinging tentacles of sea anemones. But clownfish do not get stung because they have a special coating on their skin.

The porcupine fish is covered with sharp spines. When in danger it blows itself up like a balloon to frighten the enemy.

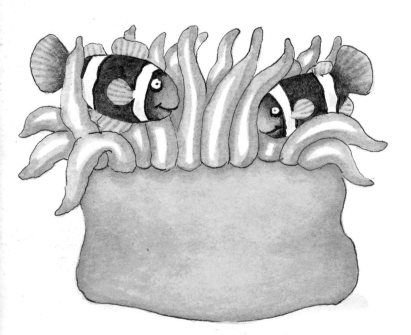

SEAHORSES

The seahorse is another very strange fish. It has a long snout with a tiny mouth at the end. The seahorse sucks up tiny sea creatures for food.

Seahorses swim upright by waving a fin on their backs. They move very slowly through the water.

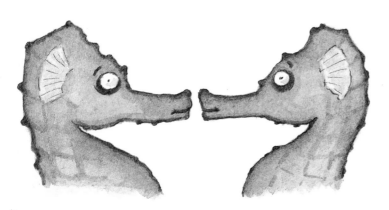

Seahorses use their curly tails to hold on to sea plants when they want to keep still.

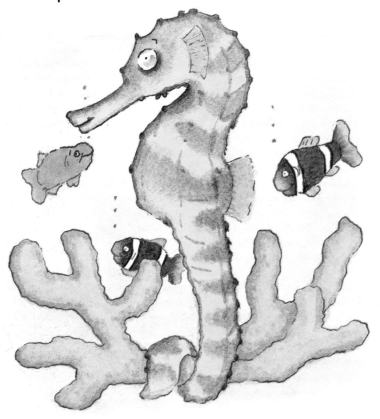

Seahorses are amazing! They can change color from yellow or orange to black in minutes

SEA TURTLES

Sea turtles come from the same family as tortoises. They have flippers instead of legs to help them swim.

When the baby turtles hatch, they scamper down to the sea. They have to go very fast because hungry sea birds are waiting to catch them.

Female turtles lay their eggs on sandy beaches. They dig a hole and cover the eggs with sand.

When they are frightened, most turtles pull their head and flippers into their shell.

QUIZ

What do sea anemones use their tentacles for?

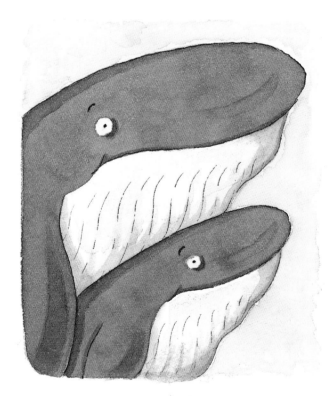

What do baby whales feed on?

What do seahorses use to help them keep still?

How do turtles swim?

How do dolphins talk to each other?

Do crabs walk forwards?

What is this strange fish called?

INDEX

crabs   14
dolphins   12
octopuses   8
quiz   22
sea anemones   10
sea turtles   20
seahorses   18
starfish   6
strange fish   16
whales   4

Published in the United States and Canada by
Two-Can Publishing LLC
234 Nassau Street
Princeton, NJ 08542

www.two-canpublishing.com

© 2002, 2000 Two-Can Publishing
Illustration © Siobhan Dodds

For information on Two-Can books and multimedia,
call 1-609-921-6700, fax 1-609-921-3349, or visit our website at
http://www.two-canpublishing.com

Created by
act-two
346 Old Street
London EC1V 9RB

'Two-Can' is a trademark of Two-Can Publishing.
Two-Can Publishing is a division of Zenith Entertainment plc,
43-45 Dorset Street, London W1U 7NA

HC ISBN 1-58728-854-0
SC ISBN 1-58728-861-3

HC 1 2 3 4 5 6 7 8 9 10 04 03 02
SC   2 3 4 5 6 7 8 9 10 04 03 02

Photo credits:p.2-3 Zefa, p.5 Britstock, p.7 Tony Stone, p.9 Bruce Coleman, p.11 Planet Earth, p.13 Ardea,
p.15 Ardea, p.17 Zefa, p.19 Zefa, p.21 Britstock

Printed in Hong Kong